Early Intervention
S KT-380-365

IMAGES

Machines

Karen Bryant-Mole

First published in Great Britain by Heinemann Library, an imprint of
Heinemann Publishers (Oxford) Ltd, Halley Court, Jordan Hill, Oxford OX2 8EJ

MADRID ATHENS PARIS FLORENCE PRAGUE WARSAW
PORTSMOUTH NH CHICAGO SAO PAULO SINGAPORE TOKYO
MELBOURNE AUCKLAND IBADAN GABORONE JOHANNESBURG

© BryantMole Books 1996

All rights reserved. No part of this publication may be reproduced, stored in a retrieval system, or transmitted in any form or by any means, electronic, mechanical, photocopying, recording, or otherwise without either the prior written permission of the Publishers or a licence permitting restricted copying in the United Kingdom issued by the Copyright Licensing Agency Ltd, 90 Tottenham Court Road, London W1P 9HE

Designed by Jean Wheeler
Commissioned photography by Zul Mukhida
Printed in Hong Kong

00 99 98 97 96
10 9 8 7 6 5 4 3 2 1

ISBN 0 431 06287 0

British Library Cataloguing in Publication Data
Bryant-Mole, Karen
Machines. – (Images Series)
I. Title II. Series
621.8

Some of the more difficult words in this book are explained in the glossary.

Acknowledgements

The Publishers would like to thank the following for permission to reproduce photographs. Chapel Studios; 15 (both), Postive Images;4 (right) Tony Stone Images; 4 (left) John Edwards, 5 (left) Zigy Kalunzy, 5 (right) Graeme Norways, 8 (left) Oli Tennent, 9 (left) R Smith, 9 (right) Lori Adamski Peek, 12 (left) Mark Wagner, 12 (right), 13 (left) John Warden, 13 (right) Baron Wolman, 14 (left) Andy Sacks, 14 (right) Barry Marsden, 16 (left) Anthony Meshkinyar, 18 (left) James Andrew Bareham, 18 (right) John Lund, 19 (left) Arnulf Husmo, 22 (right) Anthony Meshkinyar, Zefa; 8 (right),16 (right), 19 (right), 23 (left).

Every effort has been made to contact copyright holders of any material reproduced in this book. Any omissions will be rectified in subsequent printings if notice is given to the Publisher.

Contents

On the farm

These machines help farmers in their work.

Music

Do you like listening to music?

FM
MW
88 92 96 100 104 106 108
54 60 70 80 100 120 140 160
MHz
x10 kHz
SRX 65
FM/MW STEREO RADIO RECORDER
TUNING
10 11 11+ 12 12+ 13 13+ 14 15

On the move

These machines take people from place to place.

Which of these machines have you used?

Toys

These machines can be played with.

You can see the working parts inside this helicopter.

Flying

a plane

a hang glider

a hot air balloon

a helicopter

Which machine do you think moves the fastest?

Building sites

These huge machines lift, carry, mix and dig.

Clean and tidy

Machines like these keep us clean and tidy.

They are all powered by electricity.

At sea

These machines carry people and goods over the sea.

RESCUE

Tiny machines

Some machines
are very small.

There is a tiny radio inside this hat!

At home

There are lots of machines in our homes.

Can you think of any more?

Glossary

goods objects or things
machine an object with moving parts
powered what makes an object work

Index